Dark Energy, Dark Matter and Anti-Gravity

Hidden in Plain Sight (Hot Science)

BALUNGI FRANCIS

TABLE OF CONTENTS

DEDICATION

To my wife W. Ritah for his constant feedback throughout and many long hours of editing,

To my sons Odhran and Leander,

To Carlo Rovelli, Lee Smolin, Neil deGrasse Tyson and Sabine Hossenfielder, I say thank you for your astonishing suggestions.

PREFACE

The universe is full of matter and the attractive force of gravity pulls all matter together.

The expansion of the universe has not been slowing due to gravity, as everyone thought, it has been accelerating.

Maybe it was a result of a long-discarded version of Einstein's theory of gravity, one that contained what was called a "cosmological constant." Maybe there was some strange kind of energy-fluid that filled space. Maybe there is something wrong with Einstein's theory of gravity and a new theory could include some kind of field that creates this cosmic acceleration.

It turns out that roughly 68% of the universe is dark energy. Dark matter makes up about 27%. The rest - everything on Earth, everything ever observed with all of our instruments, all normal matter - adds up to less than 5% of the universe. Come to think of it, maybe it shouldn't be called "normal" matter at all, since it is such a small fraction of the universe.

Balungi Francis explains this major conundrum in modern science and looks at how scientists are beginning to find solutions to it.

PART I

DARK ENERGY

1. EXPANDING UNIVERSE

Dark energy is a hypothetical form of energy that affects the universe on the largest scales. It's thought to be the cause of the accelerating expansion of the universe.

Dark energy is an "anti-gravity" force that fills the universe and stretches the fabric of spacetime. It drives cosmic objects apart at an increasingly rapid rate rather than drawing them together as gravity does.

Dark energy is thought to be present in such a large quantity that it overwhelms all other components of matter and energy put together.

Dark energy was discovered in 1998 by two international teams that included American astronomers Adam Riess and Saul Perlmutter and Australian astronomer Brian Schmidt.

Why is dark energy repulsive to gravity?

Dark energy is repulsive to gravity because it has a negative pressure and a constant energy density. The pressure is greater than the energy density, which results in a repulsive force.

Dark energy's negative internal pressure causes repulsive gravity. Dark energy's even distribution means it doesn't have any local gravitational effects, but rather a global effect on the universe. This leads to a repulsive force that accelerates the expansion of the universe.

Does dark energy interact with gravity?

Dark energy is thought to interact with gravity, but not with any other fundamental forces.

Dark energy is a mysterious force that causes the universe to expand at an accelerating rate. This acceleration is often described as "anti-gravity" because it appears to counteract the attractive force of gravity.

Dark energy is distributed evenly throughout the universe, so its effect is not diluted as the universe expands. This means that dark energy doesn't have any local gravitational effects, but rather a global effect on the universe as a whole.

What is the physics behind dark energy?

In Einstein's theory of general relativity, dark energy is a component of the stress-energy tensor in the field equations. In this theory, the shape of space-time corresponds to the matter-energy of the universe.

Dark energy is a property of space, even empty space. It's not a material that comes from somewhere. As space expands, there's more dark energy because there's more volume.

Dark energy is thought to be the result of a quantum disentanglement physical process.
One idea is that dark energy is a fifth fundamental force called quintessence, which fills the universe like a fluid.

Another explanation is that dark energy is a new kind of dynamical energy fluid or field. It fills all of space, but its effect on the expansion of the universe is the opposite of that of matter and normal energy.

What kind of force is dark energy?

Dark energy is a scalar field, which means it has no direction. It's a repulsive force that causes space-time to expand.

Dark energy is the dominant component of the universe, making up 69.4% of it. The remaining portion is made up of ordinary matter and dark matter.

Dark energy is a mysterious force that causes the universe to expand at an accelerating rate. This is contrary to what one might expect from a universe that began in a Big Bang.

Dark energy's density is very low, much less than the density of ordinary matter or dark matter within galaxies. However, it dominates the universe's mass–energy content because it is uniform across space.

Some scientists believe that dark energy is a fifth fundamental force called quintessence, which fills the universe like a fluid. Others say that the known properties of dark energy are consistent with a cosmological constant.

What is the equation for dark energy?

In the literature, dark energy is usually described by $w \equiv P/\varrho$, where P and ϱ denote its pressure and energy density.

Therefore, exploring the evolution of w is the key approach to understanding dark energy.

Is dark energy weaker than gravity?

Some say it's not meaningful to compare dark energy and gravity in terms of strength. Dark energy is extremely weak but constant, while gravity can be very strong or even weaker than dark energy.

In areas with lots of matter, gravity's attractive forces are greater than dark energy's repulsive forces. In mostly empty space, dark energy's repulsive forces are much larger than gravity's attractive forces.

Gravity is considered weak compared to other fundamental forces, like electromagnetism and the strong and weak nuclear forces. One reason for this is that gravity's effects are cumulative but are easily overshadowed by other forces at smaller scales.

Dark energy is anti-gravity, but only for space itself. It causes space to repel itself.

Is dark energy just anti-gravity?

Think of dark energy as the "evil counterpart" to gravity–an "anti-gravity" force providing a negative pressure that fills the universe and stretches the very fabric of spacetime. As it does so dark energy drives cosmic objects apart at an increasingly rapid rate rather than drawing them together as gravity does.

When did dark energy overtake gravity?

Dark energy's acceleration likely overtook gravity's pull about 5 billion years ago. Astronomers believe dark energy's repulsive force overtook gravity's attractive force about 5–6 billion years ago.

Dark energy is thought to have become the dominant force in the universe around 5 billion years ago, and to have started to really dominate around 7 billion years ago.

Will dark energy overcome gravity?

Dark energy overpowers gravity and is expanding the universe.

Dark energy is a mysterious force that is causing the expansion of the universe to accelerate. It works against the force of gravity to drive matter in the universe farther apart.

Dark energy's repulsion will never overcome the attractive forces inside established systems if its pressure is exactly the negative of its energy density.

Gravity still beats dark energy at small scales.

Why is dark energy repulsive to gravity?

Dark energy is a repulsive force to gravity because it has a negative pressure and a constant energy density. The negative pressure is a source of gravity, and since it's greater than the energy density, it results in a repulsive nature.

Dark energy is sometimes called "antigravity" because it's a term in gravity with the opposite sign. It's also considered the "evil counterpart" to gravity, filling the universe with a negative pressure that stretches the fabric of space-time.

Dark energy has a global effect on the universe, rather than local gravitational effects. This leads to a repulsive force that accelerates the expansion of the universe.

Is there more dark energy than matter?

According to the standard lambda-CDM model of cosmology, the universe is made up of:

1. 5% ordinary matter
2. 26.8% dark matter
3. 68.2% dark energy

This means that dark energy is the dominant component of the universe, contributing 68% of the total energy. Dark matter makes up about 27% of the universe, and the rest is less than 5%.

Dark energy is distributed evenly throughout the universe, not only in space but also in time. This means that its effect is not diluted as the universe expands.

Does dark energy have mass?

Dark energy is thought to be massless, with a density of roughly 10–27 kg/m3. It's also not known to interact with any fundamental forces other than gravity. Because of this, it's unlikely to be detectable in laboratory experiments.

Dark energy is thought to be made up of photons with negative inertial and gravitational mass. However, these photons would be unobservable, even though they exert a repulsive gravitational force on photons with positive mass.

Dark energy is not believed to have any material existence, unlike dark matter. It's more of an anti-gravitational field, first conceived by Einstein as his "cosmological constant".

Does dark energy have a physical form?

The exact nature of dark energy is still a mystery. However, some proposed forms of dark energy include:

4. **Cosmological constant:** A constant energy density that fills space homogeneously
5. **Scalar fields**: Dynamic quantities whose energy density can vary in time and space

Dark energy is thought to behave neither as a particle nor as a field, but rather as a property that is inherent to space itself. It's called "dark" because it has no electric charge and does not interact with electromagnetic radiation, such as light.

Some scientists suggest that dark energy is created and remains inside black holes, which form in the crushing forces of collapsing stars.

Does dark energy have a particle?

Dark energy is attributed to the interaction between particles that contribute to the expansion of the universe. However, there is no conclusive evidence to identify these particles.

Dark energy is thought to decay into relativistic matter, which includes particles that travel at speeds close to the speed of light. These studies provide insights into the behavior and implications of dark energy decay to relativistic matter in the context of cosmology and particle physics.

What is inside dark energy?

The exact nature of dark energy is still a mystery, but here are some theories:

6. **Pent-up energy**: Dark energy is a type of pent-up energy that's part of the fabric of space-time.
7. **Transient vacuum energy**: Dark energy is a transient vacuum energy that comes from the potential energy of a dynamical field. This form of dark energy is called "quintessence" and varies in space and time.
8. **Cosmological constant**: Dark energy is a cosmological constant, which is a constant energy density that fills space evenly.
9. **Scalar fields**: Dark energy is a scalar field, which is a dynamic quantity with energy densities that change in time and space.
10. **Vacuum energy of space**: Dark energy is the vacuum energy of space, which is when particles appear and disappear in empty space.
11. **Fifth force**: Dark energy is a "fifth force" that's responsible for the negative pressure that may cause the universe to expand faster.

Dark energy is a mysterious force that opposes gravity and causes the universe to expand faster. About 69% of the universe is made of dark energy, and about 26% is made of dark matter. Only about 5% of the universe is made of familiar atomic matter.

Is dark energy hot or cold?

Dark matter is classified as "cold", "warm", or "hot" based on its velocity.

Dark matter is considered "cold" if its particles move slowly compared to the speed of light. Lighter, faster particles are known as "hot dark matter", while heavier, slower particles are known as "cold dark matter".

Dark matter inside galaxies is extremely hot, in a cosmological sense. This is why smaller galaxies cannot form by gravitational clumping inside larger galaxies.

Dark matter does not take part in the electromagnetic interaction, so it neither absorbs nor emits electromagnetic waves. This means that if dark matter particles do have a "temperature", then we do not have any direct means of measuring it.

What does dark energy really do?

Dark energy's primary effect is to drive the expansion of the universe. Dark energy is a "anti-gravity" force that provides a negative pressure that fills the universe and stretches the fabric of spacetime. As it does so, dark energy drives cosmic objects apart at an increasingly rapid rate.

Dark energy works against gravity, so more dark energy accelerates the universe's expansion and retards the formation of large-scale structure.

One technique for measuring the expansion rate is to observe the apparent brightness of objects of known luminosity like Type Ia supernovas.

Dark energy is a property of space itself, so it would not be diluted as space expands. As more space comes into existence, more of this energy-of-space would appear. As a result, this form of energy would cause the universe to expand faster and faster.

While there are various speculative ideas about how dark energy might be harnessed or manipulated, there is currently no experimental evidence or technology to support these concepts.

Dark energy vs Dark matter

Dark matter and dark energy have similar names but opposite effects:

12. **Dark energy:** This hypothetical form of energy exerts a negative, repulsive pressure that counteracts gravity. Dark energy is an intrinsic property of space, so it's not diluted by the expansion of space. Dark energy acts on the universe as a whole.
13. **Dark matter:** This type of matter is responsible for gravity and keeps stars, planets, and moons in orbit. Dark matter is non-baryonic and non-luminous, meaning it doesn't emit, reflect, or absorb light or electromagnetic radiation. Dark matter is also non-interacting with the electromagnetic force, making it difficult to detect. Dark matter affects individual galaxies.

Dark matter slows down the expansion of the universe, while dark energy speeds it up.

Can we control dark energy?

Dark energy can be controlled because it's part of the composition of universes. This user also says that dark energy gives the power of reality manipulation.

However, dark energy's properties are not fully understood. Dark energy may destroy itself and decay, or it may convert into baryonic matter or a new particle. Because we don't understand how it's made, we also don't know how to destroy it.

If we could harness the power of dark energy, it could potentially lead to revolutionary advancements in space travel, energy generation, and understanding the fundamental forces of the universe.

Is dark energy conserved?

Dark energy is conserved in a volume that remains constant, meaning it doesn't expand with the universe.

However, the conservation of energy doesn't apply universally in cosmology. The universe isn't considered an isolated system, and its expansion doesn't follow the same rules as typical isolated energy systems.

As the universe expands, its total energy isn't conserved due to the presence and behavior of dark energy. The energy density of dark energy is constant as the universe expands. However, the energy densities of other forms of matter, such as radiation and dark matter, decreases with an expanding universe.

Can we extract dark energy?

It's not clear if dark energy can be extracted or harnessed for use. Dark energy's properties and potential applications are still largely unknown.

In theory, dark energy could be used as a power source. However, in practice, the amount of energy that could be liberated is too small to be useful or detectable.

Dark energy comes in the form of pulling things apart as the universe expands. Any engine that utilizes this would need to expand indefinitely with the universe while still remaining in one piece.

Dark energy can be obtained from the interplay of the Higgs boson and the inflaton. It can also be extracted through the vacuum condensate induced by neutrino mixing phenomenon.

Is dark energy an energy source?

Dark energy is a hypothetical form of energy that affects the universe on the largest scales. It is the dominant component of the universe, contributing 68% of the total energy.

Dark energy is an unknown energy source that causes the universe's expansion to accelerate. It has a very low density, but it dominates the universe's mass–energy content because it is uniform across space.

In theory, it is possible to use dark energy as a power source. However, the amount of energy that could be liberated in a local setting is too small to be useful or even detectable.

Dark energy was discovered in 1998 by two international teams that included American astronomers Adam Riess, Saul Perlmutter, and Australian astronomer Brian Schmidt.

Does dark energy have a physical form?

Dark energy is a hypothetical form of energy that pervades space and has a negative pressure. It's the most accepted hypothesis to explain the observations since the 1990s that the universe is expanding at an accelerating rate.

Dark energy is called "dark" because it has no electric charge and does not interact with electromagnetic radiation, such as light.

Dark energy is not directly observed. It behaves neither as a particle nor as a field, but rather as a property that is inherent to space itself.

Is dark energy just gravity?

Dark energy is the opposite of gravity. It's a force that pushes cosmic objects apart, while gravity pulls them together. Dark energy is also known as an "anti-gravity" force.

Dark energy is distributed evenly throughout the universe, so it has a global effect on the universe as a whole. This means that dark energy doesn't have any local gravitational effects.

Dark energy accounts for around 68% of the universe's energy and matter content. Somewhere between 3 and 7 billion years after the Big Bang, dark energy became more influential than gravity and the expansion of the universe has been accelerating ever since.

Is dark energy quantum physics?

Some say that dark energy is a natural consequence of a quantum disentanglement physical process. Others say that dark energy is a vacuum energy of the sort that quantum particles might create by popping in and out of free space.

Dark energy could also be related to the cosmological constant, a term in Einstein's equations of general relativity. If this is true, dark energy could be associated with the energy of the vacuum itself according to Quantum field Theory.

However, when physicists try to calculate the energy of the vacuum using Quantum field Theory, they get a number that's way too large. This is one of the biggest unsolved problems in theoretical physics, known as the cosmological constant problem.

Does dark energy obey the laws of physics?

There's no evidence that dark energy doesn't follow the rules of physics. However, some say that dark energy violates the law of energy conservation.

The law of energy conservation states that energy can't be created or destroyed, but can be converted from one form to another. This means that a system always has the same amount of energy, unless it's added from the outside.

However, some say that energy isn't conserved on the scale of the universe because the universe's expansion violates time translation symmetry.

Dark energy doesn't violate energy conservation because general relativity uses a different conservation law. This law is the local conservation of all kinds of energy densities, and dark energy fulfills this law.

Dark energy is more hypothetical than dark matter, and many things about it remain speculative. Dark energy is thought to be very homogeneous and not dense, and is not known to interact through any of the fundamental forces other than gravity.

Is dark energy a force or energy?

Dark energy is both a force and a form of energy. It's a mysterious force that causes the universe to expand at an accelerating rate.

Scientists believe dark energy is a property of space itself or a new type of energy field. Some say it's a fifth fundamental force called quintessence that fills the universe like a fluid.

What is the main source of dark energy?

According to some scientists, dark energy is produced by a combination of black holes and the expansion of the universe.

The cosmological constant is the most likely candidate for dark energy. This is often related to the fluctuations of the quantum vacuum.

Other sources of dark energy include:

14. Neutrinos
15. Symmetric neutrino-like particles
16. Sterile neutrinos

Dark energy is inherent to the expansion of the universe and remains constant. It's distributed evenly throughout the universe, not only in space but also in time.

How does dark energy affect the galaxy?

Dark energy pushes galaxies apart, while gravity pulls them together. This creates a cosmic tug of war between the two forces, and the universe's expansion or contraction depends on which force is stronger.

Dark energy's repulsive force slows down the accretion and merging of galaxies. This gradually starves galaxies of the fuel they need to form stars.

If dark energy increases, the universe's expansion may accelerate so quickly that galaxies, stars, and atoms will be torn apart. This is known as the Big Rip. Dark energy may also lead to a re-collapse of the universe, known as the Big Crunch.

How does dark energy push galaxies apart?

Dark energy pushes galaxies apart because it's a repulsive force that repels matter that comes close to it. Dark energy's constant density makes it dominant over time, even as the universe expands.

Dark energy and dark matter work together to maintain the shape and expansion of the universe and galaxies. Dark matter pulls galaxies together, while dark energy pushes them apart.

If dark energy becomes stronger over time, the universe could tear itself apart, resulting in a "Big Rip" scenario. This would tear apart all bound structures, including galaxies, stars, and even atoms.

Are galaxies held together by the force of dark energy gravity?

Gravity is the force that holds galaxies together. Gravity is an attractive force that pulls all matter in a galaxy towards its center. This keeps the stars, planets, and other celestial objects in orbit.

Dark energy is a theory that proposes that the universe is expanding due to a repulsive force. Dark energy is an "anti-gravity" force that stretches the fabric of spacetime and drives cosmic objects apart.

Scientists say that dark matter overcomes dark energy inside galaxies, so gravity wins and keeps galaxies together. However, outside of galaxies, dark energy overcomes dark matter and wins, so galaxies are moving away.

PART II

DARK MATTER

2. THE GALAXY ROTATION PROBLEM

Dark matter is a hypothetical form of matter that doesn't interact with light or the electromagnetic field. It's made up of particles that don't absorb, reflect, or emit light, so they can't be detected by observing electromagnetic radiation.

Dark matter is a component of the universe that's inferred from its gravitational attraction rather than its luminosity. It makes up 30.1% of the matter-energy composition of the universe.

Dark matter is hypothesized to explain phenomena including gravitational lensing and galactic rotation curves. For much of the time since the Big Bang, dark matter has governed the cosmos, forming galaxies and galaxy clusters. Dark matter particles called "macros" could stream through space and constantly bombard Earth. Some could seriously injure any unlucky humans they pass through, but a lack of mysterious deaths suggests the biggest potential macros don't exist.

Since the 1970s and early 1980s, a growing amount of observational data has been accumulating that shows that Newtonian and Einstein gravity cannot describe the motion of the outermost stars and gas in galaxies correctly, if only their visible mass is accounted for in the gravitational field equations.

To save Einstein's and Newton's theories, many physicists and astronomers have postulated that there must exist a large amount of "dark matter" in galaxies and also clusters of galaxies that could strengthen the pull of gravity and lead to an agreement of the theories with the data. This invisible and undetected matter removes any need to modify Newton's and Einstein's gravitational theories. Invoking dark matter is a less radical, less scary alternative for most physicists than inventing a new theory of gravity.

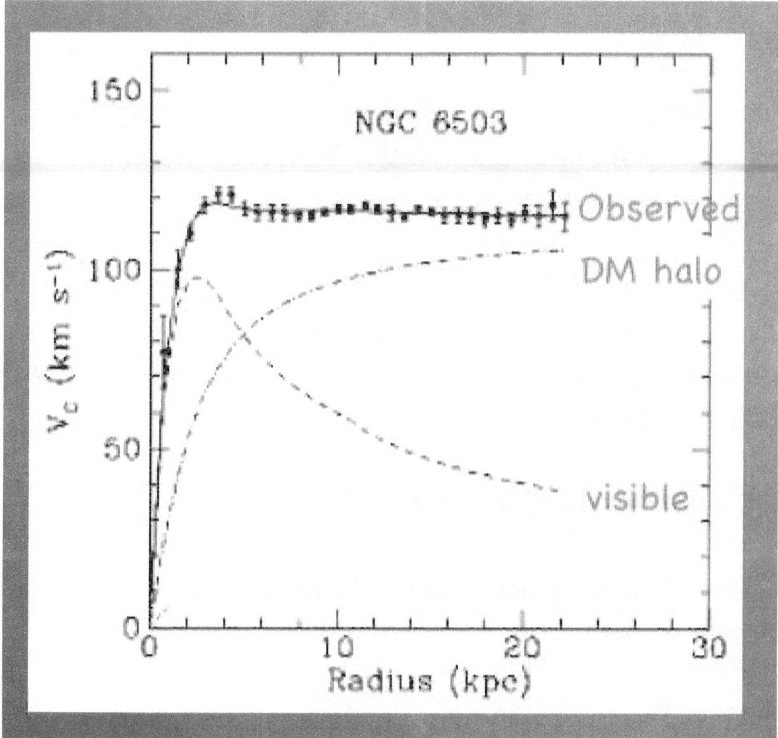

Fig. Galaxy data that show that Newtonian and Einstein gravity do not fit the observed speed of stars in orbits inside a galaxy such as NGC 6503

If dark matter is not detected and does not exist, then Einstein's and Newton's gravity theories must be modified.

Can this be done successfully? Yes! My modified gravity (MOG) can explain the astrophysical, astronomical and cosmological data without dark matter.

Why dark matter is called dark?

Dark matter is called "dark" because it can't be seen with the human eye or any optical instruments. It's also called "dark" because it symbolizes the scientific community's lack of understanding about its nature, properties, and behaviors.

Dark matter is also called "matter" because it has a gravitational effect like matter. However, it's called "dark" because it's an unknown mystery to us.

Dark matter is invisible, and light of all types seems to pass through it as though it's completely transparent. However, dark matter does have mass, which we see by its gravitational influence.

Does dark matter have gravity?

Yes, dark matter has gravity. In fact, its gravitational pull may be greater than standard matter.

Dark matter's gravitational pull helps keep galaxies from flying apart and stars from swirling out into space. Dark matter's gravity can also bend light traveling from distant galaxies, causing their images to appear distorted.

Dark matter's gravitational influence can be detected by observing how it bends and distorts light from more-distant objects. This phenomenon is called gravitational lensing.

How does dark matter react with gravity?

Dark matter is believed to contribute to the gravitational pull of galaxies and galaxy clusters. Dark matter is invisible material that adds to the gravitation caused by visible matter in or around galaxies.

Dark matter interacts with normal matter gravitationally. This is how scientists can infer that it exists. Dark matter is only inferred from its gravitational effects because it doesn't interact with the electromagnetic force and doesn't absorb or emit light.

Dark matter particles can occasionally annihilate each other through chance gravitational interactions. These chance interactions are known as "gravity portals".

The strength of gravity determines the size and distribution of dark matter structures. The size and distribution in turn determine how warped galaxies appear.

Dark matter particles can penetrate all other forms of matter. This means that they may even be able to traverse right through our planet without losing any energy. However, their impact with ordinary matter may hamper them slightly, resulting in a loss of energy.

Is the origin of dark matter gravity itself?

The existence of dark matter is a manifestation of gravity itself.

Theorists believe that dark matter was created in the aftermath of the Big Bang. Researchers have only been able to infer the existence of dark matter from the gravitational effects it has on visible matter.

The dark matter effect is caused by the negative gravity of dark energy. Dark energy is considered an "anti-gravity" force that stretches the fabric of spacetime and drives cosmic objects apart.

Is dark matter gravity from another dimension?

Some physicists believe that dark matter could be gravity from another dimension.

Here are some theories about dark matter and other dimensions:

17. **Dark matter is 4D:** If dark matter is 4D and interacts with gravity, then gravity would have to be a 4D field.
18. **Dark matter is the gravitation from another universe:** Dark matter could be the gravitation from matter in another nearby universe. This would explain why we don't see it and why it doesn't bump into things.
19. **Dark matter is an extra dimension**

Is dark matter black holes?

Dark matter would appear as a continuous sequence of adjacent particles within this proposed fourth dimension.

This would explain how dark matter causes some of the curious behaviors seen in small galaxies.

As of 2020, dark matter is only a theoretical substance. However, there are two main theories about what dark matter is:

20. **Subatomic particles:** Dark matter could be made up of undiscovered subatomic particles, such as axions or weakly interacting massive particles (WIMPs).
21. **Primordial black holes:** Dark matter could be made up of primordial black holes (PBHs) that formed after the Big Bang.

Some researchers have become more open to the idea of dark matter being made up of PBHs. However, as of 2023, evidence that PBHs may be dark matter is inconclusive.

One reason why dark matter may not be made up of black holes is that black holes are discrete entities of warped space-time. To produce the smooth gravitational effects of dark matter, enormous numbers of small black holes would need to be evenly distributed across vast regions. However, a "soup" of black holes would be unstable.

Can a black hole swallow dark matter?

Yes, black holes can swallow dark matter that comes close enough to them.

According to a late 20th century hypothesis, dark matter that's close enough to a black hole will be swallowed, leaving remnants to be redistributed.

Dark matter behaves differently than normal matter when it's swallowed by a black hole. For example, normal matter heats up and radiates light as it falls towards a black hole.

Dark matter and black holes can't interact by the strong force that binds atomic nuclei together. Until recently, scientists believed that dark matter might interact via the weak force, but new observations have disproved this.

What can destroy dark matter?

Dark matter is a fundamental particle that cannot be destroyed by physical processes. However, NBC News says that if dark matter is made up of neutralinos, then dark matter particles can self-destruct when they collide.

Dark matter particles can penetrate all other forms of matter, including Earth. However, their impact with ordinary matter may slightly hamper them, resulting in a loss of energy.

Can light destroy dark matter?

There is no known way to destroy dark matter. Dark matter interacts gravitationally with ordinary matter and itself, but it doesn't undergo processes of destruction in the same way that ordinary matter does.

Dark matter doesn't interact with light or electromagnetic forces because it's not made of particles that have an electric charge. Most matter we know of, like atoms and molecules, interacts with light because of the electric charge carried by their constituent particles.

Dark matter is also invisible and undetectable through telescopes. It's currently impossible for humans to touch or interact with dark matter in any direct way.

Why doesn't light interact with dark matter?

Dark matter doesn't interact with light because it doesn't interact with the electromagnetic field. The electromagnetic field is required for something to be affected by photons, except for their gravitational effect on space time.

Dark matter doesn't absorb, reflect, or emit light, making it difficult to detect. Researchers have only been able to infer the existence of dark matter from the gravitational effect it seems to have on visible matter.
Dark matter does interact with regular, or baryonic, matter via gravity. It's possible that particles of dark matter may occasionally smash into particles of baryonic matter.

Dark matter doesn't experience the electromagnetic interaction and so can't clump together. Electromagnetic interactions between charges are what form atoms and what allows atoms to bond together to form molecules

Is dark matter the opposite of light?

The concept of "opposite" isn't well-defined in science. However, dark matter is called "dark" because it doesn't interact with light. Dark matter is transparent and colorless.

Dark matter is a hypothetical form of matter that doesn't interact with light or the electromagnetic field. It's called "dark" because it doesn't absorb or emit electromagnetic radiation (light). Dark matter can interact with light via gravitational lensing.

Dark matter is different from antimatter, which is the opposite of normal matter. Dark matter might have an anti-version of itself, or it might consist of particles that are their own antiparticles.

Can light escape dark matter?

Light can't travel through dark matter in the same way it travels through regular matter. However, dark matter's gravitational effects can influence the path that light takes as it travels through the universe.

Light rays that pass close to a black hole get caught and cannot escape. The region around the black hole is a dark disk. Light rays that pass a little further away don't get caught but do get bent by the black hole's gravity.

Once a particle of light ('photon') passes the 'event horizon' of a black hole, it can no longer escape. However, there's nothing to suggest that it is destroyed.

Light that is near a black hole, but not actually inside it, can certainly escape away to the rest of the universe. This effect is in fact what enables us to indirectly "see" black holes.

What happens if dark matter touches Earth?

Because dark matter doesn't interact with matter, it would pass through the Earth as if it were empty space.

Dark matter particles can build up in the Earth's core. If the density of dark matter is high enough, the particles will eventually annihilate each other. This process adds a large amount of internal heat to the Earth, which can cause global pulses of geologic activity.

Dark matter particles have a very low probability of interacting with atoms inside the Earth. However, underground detectors are tuned to capture rare events of dark matter particle collisions with an atom inside a detector.

Does dark matter interact with Earth?

Some scientists believe that dark matter may exist on Earth, but there's no way to test this.

Dark matter may create a halo around Earth and other objects made of normal matter. One astrophysicist estimates that 24 trillion metric tons of dark matter surround Earth in the space between the Earth and the moon.

Dark matter can also interact with the protons and neutrons in the nucleus of atoms, producing flashes of light called scattering.

Dark matter's gravity drives normal matter (gas and dust) to collect and build up into stars and galaxies. Scientists refer to it as the glue which holds our solar system, galaxies, and clusters of galaxies together.

How does dark matter affect the world?

Dark matter is thought to affect the universe in several ways, including:

22. **Gravity:** Dark matter's gravitational pull is believed to play a key role in the formation and structure of galaxies, as well as in the large-scale distribution of matter in the universe. Dark matter also provides the extra gravity that keeps stars from swirling out into space.
23. **Cosmic web:** Dark matter's properties affect how the Universe evolved, like how galaxies form and grow as well as how they clump together to form the largest structure in the Universe: the cosmic web.
24. **Density and velocity:** Dark matter affects the cosmic microwave background (CMB) by its gravitational potential and by its effects on the density and velocity of ordinary matter.
25. **Galaxy rotation:** According to cosmologists' prevailing theory, dark matter pervades pretty much every galaxy, providing the extra gravity that keeps stars from swirling out into space.
26. **Galaxy mass:** Without dark matter, galaxies don't have enough mass to stay together—they would fly apart.

Dark matter is also crucial for all aspects of life and reality as we know them. Dark matter keeps us in the sun's orbit and keeps the sun in the Milky Way's orbit.

How does dark matter affect gravity?

Dark matter has a gravitational pull that's greater than standard matter. Dark matter's gravity is thought to play a key role in the formation and structure of galaxies. Dark matter's gravity also keeps stars from flying apart and gives galaxies their spin and mass.

Does dark matter emit gravitational waves?

Scientists believe that dark matter could generate gravitational waves strong enough for equipment like LIGO to detect.

Gravitational waves are waves in spacetime that interact with both visible and dark matter in the same way. This is different from electromagnetic interactions, which don't interact with dark matter.

The strongest gravitational waves come from cataclysmic events like colliding black holes, supernovae, and colliding neutron stars.

Does dark matter emit anything?

Dark matter doesn't emit any type of electromagnetic radiation, including infrared radiation, radio waves, ultraviolet radiation, X-rays, or gamma rays. It's completely invisible and doesn't emit light or energy.

However, some theories predict that dark matter particles may sometimes annihilate with each other to produce gamma rays. This would result in a weak gamma-ray signal from parts of the galaxy that are otherwise too "boring" to produce such energetic radiation.

Does dark matter make dark energy?

The decline of dark matter relative to dark energy doesn't mean that dark matter is decaying or transforming into dark energy.

Theortical physicists and observational astronomers generally expect dark matter and dark energy to be different entities, and most research is based on the idea that they have little relationship to one another.

Dark matter is a hypothetical form of unseen matter that makes up 85% of the universe's total mass. Dark energy is a mysterious form of energy that makes up 68.2% of the universe's total energy. Dark energy is a force or energy that can counteract gravity at very large scales.

Dark matter is assumed to explain gravitational behavior observed in galaxies. Dark energy is assumed to exist to explain the accelerating expansion of the universe.

How much energy does dark matter have?

Dark matter makes up 30% of the universe's total mass and energy. Dark energy is estimated to account for between 68% and 72% of the universe's total energy and matter.

Dark matter impacts individual galaxies, while dark energy acts on the universe as a whole. Dark energy is clearly more powerful as it is able to overcome attractive forces to cause accelerated expansion.

Does dark matter have infinite energy?

When dark matter annihilates, it produces pure energy. However, the current scientific understanding is that dark matter is not infinite in quantity.

Dark matter makes up 85% of the universe's total mass.

Dark matter is distributed throughout the universe, but the exact amount is still a subject of ongoing research and debate.

According to current understanding, the total energy in the universe is believed to be constant, but the distribution and availability of energy can vary.

What is the mystery of dark matter?

Dark matter is a hypothetical form of matter that doesn't seem to interact with light or the electromagnetic field. It's implied by gravitational effects that can't be explained by general relativity unless more matter is present than can be seen.

Dark matter is thought to be non-baryonic in nature, which means it's possibly made up of some undiscovered subatomic particles. These unknown particles are expected to be either weakly-interacting massive particles (WIMPs), or gravitationally-interacting massive particles (GIMPs).

Dark matter makes up over 80% of all matter in the universe, but scientists have never seen it. We only assume it exists because, without it, the behavior of stars, planets, and galaxies simply wouldn't make sense.

The problem with the dark matter hypothesis is that nobody knows what form dark matter takes. When the term was first proposed back in 1933 by the Swiss-American astronomer Fritz Zwicky, it was possible that the extra mass was simply clouds of hydrogen gas.

What are 3 facts about dark matter?

Here are some facts about dark matter:

27. Dark matter is not going away
28. Dark matter doesn't interact much with itself or anything else
29. Dark matter is believed to be non-baryonic and non-relativistic
30. Dark matter is believed to be made up of as yet undiscovered particles such as Weakly Interacting Massive Particles (WIMPs):
31. Dark matter is believed to have been "cold" even in the very early stages of the Universe, as opposed to hot or warm
32. Dark matter has a much broader distribution of velocities, both in magnitude and direction
33. Dark matter has no coherent motion, instead moving in random directions with a wide spread of speeds
34. Dark matter means the time dilation of the hierarchical structure of the universe
35. Dark matter is a theoretical type of matter that makes up about 23% of the universe. It was initially discovered as a "missing mass" that could not be accounted for by visible matter.

36. Dark matter is believed to be non-baryonic and non-relativistic, and it dominates in gravitation.
37. Dark matter is categorized into three types: hot, warm, and cold. Light, fast particles are known as hot dark matter; heavy, slow ones are cold dark matter; and warm dark matter falls in between.

Why is dark matter a problem?

The key problem with the dark matter hypothesis is that nobody knows what form dark matter might take. Despite recent advances in astrophysics and astronomy, scientists still don't understand exactly how galaxies can exist.

Can we use dark matter as energy?

Dark matter is too inert to be used as fuel. Dark matter interacts so little with ordinary matter that it would spill out of a container.

However, some say that dark matter could be an unlimited fuel source with 100% efficient matter-to-energy conversion potential. According to Futurism, if dark matter could be focused, it could provide an unlimited supply of energy.

In theory, dark energy could be used as a power source. However, in practice, the amount of energy that could be liberated is too small to be useful or detectable.

How did dark matter contribute to the formation of galaxies?

Dark matter is thought to have played a major role in the formation of galaxies and larger structures like clusters of galaxies.

Here's how dark matter is thought to have contributed to the formation of galaxies:

38. **Gravitational force:** Dark matter provides the gravitational force to pull elements of the universe together to form stars. It can also draw other matter in the universe towards it and hold entire galaxies.
39. **Gravitational foundation:** Dark matter is thought to provide a scaffolding of sorts, creating a gravitational foundation that allows gas and dust to come together and form stars.
40. **Counteract expansion of the universe:** Dark matter provided the additional gravitational force needed to counteract the expansion of the universe and allow structures like galaxies and clusters of galaxies to form.
41. **Form protogalactic clouds:** Because most of a galaxy's mass is in the form of dark matter, the gravity of that dark matter is probably what formed protogalactic clouds and then galaxies from slight density enhancements in the early universe.

Dark matter is an invisible form of matter whose total mass in the universe is roughly five times that of "normal" matter (i.e., atoms). All galaxies are made of mostly dark matter. The visible matter, like stars, gas and dust, makes up only a small percentage of a galaxy's mass.

Is dark matter necessary to explain the gravitational force holding galaxies together?

Dark matter is thought to be necessary to form galaxies as we observe them. Without dark matter, galaxies would form, exist, and live stably just fine. However, they would look quite different than the galaxies we actually see.

Dark matter is thought to be the "glue" that holds galaxies together. Astronomers suggest that dark matter provides vital "scaffolding" for the universe, forming a framework for the formation of galaxies through gravitational attraction.

Dark matter's gravity drives normal matter (gas and dust) to collect and build up into stars and galaxies. Scientists simply refer to it as the glue which holds our solar system, galaxies, and clusters of galaxies together.

The main way we know dark matter exists is from its gravitational effect of holding galaxies together. Early radio astronomy observations showed a half-dozen galaxies spun too fast in their outer regions, pointing to the existence of dark matter.

Does dark matter only interact with other matter through gravity?

Dark matter is thought to only interact with ordinary matter through gravity. However, it's possible that it may interact in other ways "weakly".

Dark matter is a hypothetical form of matter that doesn't seem to interact with light at all. It's thought to be non-electromagnetic and non-baryonic, meaning it doesn't interact with electromagnetic forces or ordinary matter particles.

Dark matter is difficult to detect in the laboratory.

However, scientists do have overwhelming indirect evidence for dark matter.

Dark matter was first discovered decades ago because of its gravitational effect on visible matter.

We can map its presence by looking at clusters of galaxies, the most massive structures in the universe.

How does dark matter hold the universe together?

Dark matter is thought to be the glue that holds the universe together. It's believed to have coalesced before normal matter, creating gravitational wells for normal matter to fall into. These wells became the seeds of galaxies.

Dark matter's gravity creates the sponge-like structure of the universe. It also drives normal matter to collect and build up into stars and galaxies.

Dark matter makes up most of the universe's mass and creates its underlying structure. Without dark matter, the universe would look nothing like it does now. There would be no galaxies, no stars, no planets, and therefore, no life.

What is the evidence of dark matter in galaxy clusters?

Dark matter can't be seen, so astronomers use indirect evidence to detect it.

One of the most compelling pieces of evidence for dark matter is gravitational lensing. This is when the gravity of a massive object, like a cluster of galaxies, bends the light from objects behind it. The amount of bending can be used to calculate the mass of the object causing the lensing.

Another piece of evidence for dark matter is that the outlying stars in a galaxy appear to rotate more rapidly than expected.

Dark matter can also be detected by measuring the velocities of individual stars or groups of stars as they move randomly in the galaxy or as they rotate around the galaxy.

Why do we think that most of the matter in a cluster of galaxies is dark matter?

Astronomers believe that dark matter makes up most of the matter in galaxies and clusters of galaxies because it provides enough mass to keep galaxies gravitationally bound together.

Here are some reasons why astronomers believe that dark matter makes up most of the matter in galaxies and clusters of galaxies:

42. **Gravity:** Without dark matter, there wouldn't be enough gravity to hold stars in orbit around the center of the galaxy.
43. **Motion:** The mass of the visible part of a galaxy doesn't account for the motions and apparent speeds of galaxies.
44. **Orbital speeds:** The orbital speeds of stars far from the galactic center are surprisingly high, suggesting that these stars are feeling gravitational effects from unseen matter.
45. **Galaxy cluster:** In 1933, Caltech's Fritz Zwicky used the Mount Wilson Observatory to measure the visible mass of a cluster of galaxies and found that it

was much too small to prevent the galaxies from escaping the gravitational pull of the cluster.

Dark matter is a type of matter that doesn't recognize nuclear and electromagnetic forces. Instead, dark matter produces so much gravity, which allows stars in the outer orbit of a galaxy to have the same speed as those at the galactic center.

How do we determine the amount of dark matter in galaxy clusters?

Astronomers can determine the amount of dark matter in a galaxy cluster by measuring the gravitational effects of dark matter on visible objects. They can also determine the amount of dark matter in a galaxy by comparing its mass to its luminosity.

Here are some other methods used to study dark matter in galaxy clusters:

46. **Mapping distorted light:** Astronomers can deduce the mass of a cluster and the distribution of dark matter by mapping the distorted light.
47. **Comparing giant arcs:** Astronomers can compare the statistics and morphology of giant arcs in galaxy clusters using simulations.
48. **Parametric strong-lensing analysis:** Astronomers can probe the inner shape of dark matter haloes using this method.
49. **Analyzing minor merger events:** Astronomers can measure the interaction cross-section of dark matter by analyzing minor merger events.

How do you find the mass of dark matter?

The mass of dark matter in the Milky Way is about 2×10^{42} kg.

Dark matter is thought to make up about 80% of the mass in the universe. However, most of the universe is made of dark energy, which has no mass.

Dark matter can be detected indirectly through its gravitational influence on stars and galaxies. Dark matter is thought to be found wherever normal matter is.

Does dark matter have negative mass?

Dark matter is thought to have positive mass. However, some simulations using known data on dark matter suggest that dark matter could have negative mass.

Negative mass would describe instabilities, while stable configurations have positive mass. Negative mass could only "smash things together" if it created a shell around the object that pushed equally from all directions.

Dark energy, on the other hand, does have negative (relativistic) mass. Positive relativistic mass attracts, while negative relativistic mass repels. Dark energy's negative relativistic mass accelerates distant galaxies away.

What is the difference between dark matter and negative matter?

Negative matter is a hypothetical type of matter that has negative mass and negative energy. It's an invisible, exotic type of matter that has the opposite sign of mass to normal matter.

Dark matter is a mysterious form of matter that doesn't interact with the electromagnetic force. It only interacts with gravity and the weak atomic force. Dark matter is extremely hard to spot because it doesn't absorb, reflect, or emit light.

Negative matter is the simplest candidate for dark matter and can explain some characteristics of dark matter and dark energy.

What are the three types of dark matter?

Dark matter is usually divided into two categories: baryonic and non-baryonic. Non-baryonic dark matter is further divided into two classes: hot dark matter (HDM) and cold dark matter (CDM).

Dark matter can also be categorized into three types: hot, warm, and cold.

Hot dark matter: This type of dark matter is made up of light, fast particles that move at near the speed of light. Examples of hot dark matter include neutrinos.

Cold dark matter: This type of dark matter is made up of heavy, slow particles that don't move around. Examples of cold dark matter include WIMPs and hypothetical axions.

Warm dark matter: This type of dark matter falls between hot and cold dark matter. Examples of warm dark matter include sterile neutrinos.

PART III

ANTI-GRAVITY

3. ZERO-GRAVITY

Anti-gravity is a hypothetical phenomenon that involves creating a place or object that is free from the force of gravity. It also refers to reducing, canceling, or protecting against the effect of gravity.

Anti-gravity is different from the lack of weight experienced in free fall or orbit. It also differs from balancing the force of gravity with another force, like electromagnetism and aerodynamic lift.

Many scientists believe that anti-gravity is not possible, given what we know about the universe and the laws that govern it. However, some say that considering the formal analogy between gravitational theory and electromagnetic theory, it seems that anti-gravity is possible, at least theoretically.

Anti-gravity may also refer to:

Anti.Gravity: A weightless lotion that can create bigger, thicker-looking hair

Antigravity force: A force that could explain the high concentration of Chlorofluorocarbon (CFC) molecules observed in high altitudes

Is antigravity possible on Earth?

There is no technology that can neutralize the pull of gravity. According to Wikipedia, anti-gravity is impossible under general relativity, except under contrived circumstances.

However, there have been credible investigations into methods of creating an anti-gravitational force. Some say that we have never observed negative matter in the universe, and we strongly suspect it can never exist.

To escape the gravitational clutches of the Earth, you must achieve a speed of at least 25,000 mph (15,570 km/h), which is around 33 times the speed of sound.

Anti-gravity is impossible under general relativity, except under contrived circumstances.

Some say that the problem is that material with negative mass is simply not available, because no one has ever found it. Therefore, the possibility to cancel gravity using negative mass remains purely theoretical.

However, some scientific research has explored the possibility of creating conditions that could counteract the effects of gravity, such as through advanced technology or exotic forms of matter.

Is it possible to stop gravity?

According to Albert Einstein's theory of general relativity, gravity is a result of the curvature of spacetime caused by mass and energy. Gravity can't be turned off or stopped without a fundamental change in the fabric of spacetime. Gravity is a natural consequence of how space-time behaves in the presence of mass. It doesn't use up anything and doesn't need energy to operate.

There are ways to counteract the effects of gravity, such as using specialized technology like magnetic levitation or propulsion systems. However, completely stopping the force of gravity on Earth is not currently feasible.

Gravity doesn't run out because it doesn't use up anything. However, the Earth's gravitational field extends well into space and does not stop. It does weaken as one gets further from the center of the Earth.

Is it possible to go against gravity?

Humans can overcome gravity in many ways, including:
Using elevators and escalators

50. **Flying:** Parachuting and sky diving
51. **Jumping:** However, humans cannot defy gravity and fly without the help of machines. The human body is not naturally equipped for sustained flight.

Gravity can be simulated to create the sensation of weightlessness, but it cannot be completely blocked. Artificial zero gravity is possible through methods such as free-fall in space or using specialized aircraft to create brief periods of weightlessness.
You can also escape from the pull of the Earth towards its center by traveling at the Earth's escape velocity (11.2 km/s) or greater. Then you will be orbiting the earth.

Is it possible to break the laws of gravity?

No one has ever broken the law of gravity. According to Brainly.in, using rocket fuel to move against gravity doesn't break the law of gravity.

However, some say that Newton's law of gravity is an approximation that becomes less accurate in certain extreme conditions. In more intense scenarios, like around a black hole, or when more precision is needed, like when calculating GPS coordinates, you need to "break" Newton's law and upgrade to relativity.

How can you defy the laws of gravity?

There are no known objects or things that can completely negate the effects of gravity. However, there are objects and phenomena that can appear to defy gravity by counteracting its effects in specific ways.

Here are some examples of defying gravity:

52. **Magnetism:** The magnetic force between a magnet and a paperclip can be stronger than the gravitational pull of gravity. This means the paper clip remains suspended in the air rather than falling to the ground.
53. **Diamagnetism:** Small objects can be levitated by using an effect called diamagnetism.
54. **Jumping:** When you jump, your leg muscles exert force against the ground, propelling your body upwards. This action temporarily counteracts the force of gravity, allowing you to leave the ground.

Other examples of defying gravity include:
55. Skydiving
56. Bungee jumping
57. Parabolic flight
58. Amusement park rides
59. Certain acrobatic maneuvers

What force can defy gravity?

Magnetic force: However, there exists other forces of pull that can defy gravity, which is what makes us think of floating and flying objects. In this activity, we will demonstrate one such force known as magnetic force. This is commonly seen through the use of magnets, such as the ones on whiteboards or on your refrigerator.

Do anti-gravity particles exist?

According to a 2023 experiment, antigravity doesn't exist. The experiment showed that gravity pulls antihydrogen downwards, which means that antigravity doesn't exist for antimatter.

If antimatter anti-gravitates, it would be made of anti-mass or anti-energy. However, these quantities don't exist according to current understanding of physics.

Some scientific research has explored the possibility of creating conditions that could counteract the effects of gravity. For example, NASA uses aircraft to create microgravity conditions for tests and simulations.

Is there any part of space with no gravity?

There is no such thing as zero gravity in space. Gravity is everywhere in the universe, including in black holes, celestial orbits, ocean tides, and our own weight.

Gravity is a force of attraction between all objects, including stars, planets, moons, and even you. Gravity becomes weaker with distance, but it never goes completely away.

However, there is the potential for net gravitational acceleration to be equal to zero. Acceleration is a vector, and as such acceleration equal in magnitude but opposite in direction would net out to zero.

You can truly escape gravity only in deep space, beyond the domain of any planets or stars. As of yet, no technology exists to neutralize the pull of gravity.

How do the planets stay where they are if there is no gravity in space?

Gravity is the force that keeps planets in orbit around the sun. Planets are constantly falling towards the sun, but their forward motion keeps them in a stable orbit. This balance of gravitational attraction and forward motion allows the planets to move through space without any support.

Here's how gravity keeps planets in orbit:

Inertia: The momentum of planets tends to make them move in a straight line.

Gravity: Gravity generates a pulling force in the direction of the sun.

Balance: The balance between inertia and gravity determines the planet's elliptical orbit.

Without gravity, planets would continue moving in a straight line in whatever direction they were moving when gravity went away.

Why do planets have gravity but space doesn't?

Planets have gravity because they have mass. Mass is a fundamental property of matter that causes objects to attract each other. The larger the mass of an object, the stronger its gravitational pull.

In space, the effects of gravity may seem less noticeable because there are fewer large masses nearby to exert a gravitational pull. The force that gravity exerts is related to the size and distance of the object in question.

Gravity is not so obvious in space because objects tend to orbit planets instead of hitting them. Orbiting just means that an object falls towards a planet due to gravity and continually misses it.

How do things move in space without gravity?

Objects in space move due to inertia and other forces, such as gravity, momentum, and collisions.

In the absence of a significant gravitational force, objects will continue to move in a straight line at a constant speed unless acted upon by an external force.

Once an object is set in motion in space, there is very little to slow it down, so it can continue moving at high speeds for long periods of time.

Spaceships fly in space by using rocket propulsion. Rockets carry their own supply of fuel and oxidizer, which they burn to produce thrust. This thrust propels the rocket forward, allowing it to overcome the force of gravity and travel through the vacuum of space.

How would Earth move if there was no gravity?

If there was no gravity, the Earth would not be able to rotate on its axis. The Earth's crust and gravity are what give it its round shape. If gravity were to disappear, the Earth's crustal plates would begin to lift off the Earth's mantle and fly into space.

Here are some other things that would happen if there was no gravity:
60. **Everything would float:** Without gravity, there would be nothing to pull objects down towards the Earth's surface.
61. **Everything on Earth would start floating in the air.**
62. **The atmosphere would disappear:** The atmosphere would disappear into space.
63. **The moon would collide with the Earth:** The Earth would stop rotating and the moon would collide with it.

64. **The Earth would collide with the sun:** The Earth would move into a new orbit, traveling about 10% closer to the sun. Eventually, the Earth would collide with the sun and everyone would perish.
65. **The Earth would break into pieces:** The Earth's inner core would eventually burst in a lethal explosion due to intense pressure. The Earth would break into pieces that would float around space, wreaking havoc.

Is an anti-gravity drive possible?

As of now, there is no technology that can neutralize the pull of gravity.

While there are speculative theories about antigravity, such as negative mass or exotic matter, they remain theoretical and have not been demonstrated in experiments.
Anti-gravity is impossible under general relativity except under contrived circumstances.

Some say that an anti-gravity engine is probably not possible because gravity is universal. This means that the material constitution of an object makes no difference.

How does anti-gravity propulsion work?

Anti-gravity propulsion systems are based on non-reactive forces.

Here are some theories about how anti-gravity propulsion works:

66. **Vacuum manipulation:** Opposing magnetic or electric fields create a mass repelling force. This vacuum manipulation process can be used to propel a mass.

67. **Negative gravitational properties:** The vacuum between plates is less dense and less tense than the vacuum outside. The tenser vacuum outside the plates pushes the plates together due to the negative gravitational properties of the vacuum.

Anti-gravity propulsion schemes are considered the future of

What is the difference between zero gravity and anti-gravity?

Zero gravity is a state of weightlessness, while anti-gravity is a hypothetical phenomenon.
Zero gravity is when there is no apparent gravity acting on an object. It's a condition in which the effects of gravity are not felt.

Anti-gravity is the concept of negating the force of gravity. It's a hypothetical gravitational force exerted by negative mass.

Anti-gravity is often used to refer to devices that look as if they reverse gravity, even though they operate through other means. For example, lifters fly in the air by moving air with electromagnetic fields.

THE END

Key Points:

Is there more dark energy than matter?

According to the standard lambda-CDM model of cosmology, the universe is made up of:

68. 5% ordinary matter
69. 26.8% dark matter
70. 68.2% dark energy

This means that dark energy is the dominant component of the universe, contributing 68% of the total energy. Dark matter makes up about 27% of the universe, and the rest is less than 5%.

Dark energy is distributed evenly throughout the universe, not only in space but also in time. This means that its effect is not diluted as the universe expands.

Does dark energy have a physical form?

The exact nature of dark energy is still a mystery. However, some proposed forms of dark energy include:

71. **Cosmological constant:** A constant energy density that fills space homogeneously
72. **Scalar fields**: Dynamic quantities whose energy density can vary in time and space

Dark energy is thought to behave neither as a particle nor as a field, but rather as a property that is inherent to space itself. It's called "dark" because it has no electric charge and does not interact with electromagnetic radiation, such as light.

Some scientists suggest that dark energy is created and remains inside black holes, which form in the crushing forces of collapsing stars.

What is inside dark energy?

The exact nature of dark energy is still a mystery, but here are some theories:

73. **Pent-up energy:** Dark energy is a type of pent-up energy that's part of the fabric of space-time.
74. **Transient vacuum energy:** Dark energy is a transient vacuum energy that comes from the potential energy of a dynamical field. This form of dark energy is called "quintessence" and varies in space and time.
75. **Cosmological constant:** Dark energy is a cosmological constant, which is a constant energy density that fills space evenly.
76. **Scalar fields:** Dark energy is a scalar field, which is a dynamic quantity with energy densities that change in time and space.
77. **Vacuum energy of space**: Dark energy is the vacuum energy of space, which is when particles appear and disappear in empty space.
78. **Fifth force:** Dark energy is a "fifth force" that's responsible for the negative pressure that may cause the universe to expand faster.

Dark energy is a mysterious force that opposes gravity and causes the universe to expand faster. About 69% of the universe is made of dark energy, and about 26% is made of dark matter. Only about 5% of the universe is made of familiar atomic matter.

Other sources of dark energy include:

79. Neutrinos
80. Symmetric neutrino-like particles
81. Sterile neutrinos

Dark energy is inherent to the expansion of the universe and remains constant. It's distributed evenly throughout the universe, not only in space but also in time.

Is dark matter gravity from another dimension?

Some physicists believe that dark matter could be gravity from another dimension.

Here are some theories about dark matter and other dimensions:

82. **Dark matter is 4D:** If dark matter is 4D and interacts with gravity, then gravity would have to be a 4D field.
83. **Dark matter is the gravitation from another universe:** Dark matter could be the gravitation from matter in another nearby universe. This would explain why we don't see it and why it doesn't bump into things.
84. **Dark matter is an extra dimension**

As of 2020, dark matter is only a theoretical substance. However, there are two main theories about what dark matter is:

85. **Subatomic particles:** Dark matter could be made up of undiscovered subatomic particles, such as axions or weakly interacting massive particles (WIMPs).
86. **Primordial black holes:** Dark matter could be made up of primordial black holes (PBHs) that formed after the Big Bang.

Some researchers have become more open to the idea of dark matter being made up of PBHs. However, as of 2023, evidence that PBHs may be dark matter is inconclusive.

One reason why dark matter may not be made up of black holes is that black holes are discrete entities of warped space-time. To produce the smooth gravitational effects of dark matter, enormous numbers of small black holes would need to be evenly distributed across vast regions. However, a "soup" of black holes would be unstable.

How does dark matter affect the world?

Dark matter is thought to affect the universe in several ways, including:

87. **Gravity:** Dark matter's gravitational pull is believed to play a key role in the formation and structure of galaxies, as well as in the large-scale distribution of matter in the universe. Dark matter also provides the extra gravity that keeps stars from swirling out into space.

88. **Cosmic web:** Dark matter's properties affect how the Universe evolved, like how galaxies form and grow as well as how they clump together to form the largest structure in the Universe: the cosmic web.

89. **Density and velocity:** Dark matter affects the cosmic microwave background (CMB) by its gravitational potential and by its effects on the density and velocity of ordinary matter.

90. **Galaxy rotation:** According to cosmologists' prevailing theory, dark matter pervades pretty much every galaxy, providing the extra gravity that keeps stars from swirling out into space.

91. **Galaxy mass:** Without dark matter, galaxies don't have enough mass to stay together—they would fly apart.

What are 3 facts about dark matter?

Here are some facts about dark matter:

92. Dark matter is not going away
93. Dark matter doesn't interact much with itself or anything else
94. Dark matter is believed to be non-baryonic and non-relativistic
95. Dark matter is believed to be made up of as yet undiscovered particles such as Weakly Interacting Massive Particles (WIMPs):
96. Dark matter is believed to have been "cold" even in the very early stages of the Universe, as opposed to hot or warm
97. Dark matter has a much broader distribution of velocities, both in magnitude and direction

98. Dark matter has no coherent motion, instead moving in random directions with a wide spread of speeds
99. Dark matter means the time dilation of the hierarchical structure of the universe
100.   Dark matter is a theoretical type of matter that makes up about 23% of the universe. It was initially discovered as a "missing mass" that could not be accounted for by visible matter.

101.   Dark matter is believed to be non-baryonic and non-relativistic, and it dominates in gravitation.
102.   Dark matter is categorized into three types: hot, warm, and cold. Light, fast particles are known as hot dark matter; heavy, slow ones are cold dark matter; and warm dark matter falls in between.

Why do we think that most of the matter in a cluster of galaxies is dark matter?

Astronomers believe that dark matter makes up most of the matter in galaxies and clusters of galaxies because it provides enough mass to keep galaxies gravitationally bound together.

Here are some reasons why astronomers believe that dark matter makes up most of the matter in galaxies and clusters of galaxies:

103.   **Gravity:** Without dark matter, there wouldn't be enough gravity to hold stars in orbit around the center of the galaxy.

104.	**Motion:** The mass of the visible part of a galaxy doesn't account for the motions and apparent speeds of galaxies.

105.	**Orbital speeds:** The orbital speeds of stars far from the galactic center are surprisingly high, suggesting that these stars are feeling gravitational effects from unseen matter.

106.	**Galaxy cluster:** In 1933, Caltech's Fritz Zwicky used the Mount Wilson Observatory to measure the visible mass of a cluster of galaxies and found that it was much too small to prevent the galaxies from escaping the gravitational pull of the cluster.

How does anti-gravity propulsion work?

Anti-gravity propulsion systems are based on non-reactive forces.

Here are some theories about how anti-gravity propulsion works:

107.	**Vacuum manipulation:** Opposing magnetic or electric fields create a mass repelling force. This vacuum manipulation process can be used to propel a mass.

108.	**Negative gravitational properties:** The vacuum between plates is less dense and less tense than the vacuum outside. The tenser vacuum outside the plates pushes the plates together due to the negative gravitational properties of the vacuum.

Anti-gravity propulsion schemes are considered the future of